# Chemical Engineering & the Law Forum 2021



Held at the 2021 AIChE Annual Meeting

Boston, Massachusetts, USA and Online
7 - 11 November and 15 - 19 November 2021

ISBN: 978-1-7138-5717-4

**Printed from e-media with permission by:**

Curran Associates, Inc.
57 Morehouse Lane
Red Hook, NY 12571



AIChE
120 Wall Street, FL 23
New York, NY 10005-4020

Phone: (800) 242-4363
Fax: (203) 775-5177

www.aiche.org

**Additional copies of this publication are available from:**

Curran Associates, Inc.
57 Morehouse Lane
Red Hook, NY 12571 USA
Phone: 845-758-0400
Fax: 845-758-2633
Email: curran@proceedings.com
Web: www.proceedings.com

# TABLE OF CONTENTS

# 621594 Over Half a Century of Protecting the Environment through Various Policies, Administrations, Successes, & Failures; An Overview of 50+ Years of EPA Regulations and Standards

**Abstract:**

Since the establishment of the U.S. Environmental Protection Agency (EPA) on December 2nd, 1970, the Agency has implemented a vast number of regulations to benefit the American people and the environment. Over the last 51 years, the EPA and its dedicated workers have endeavored to fulfil its mission to protect human health and the environment; the Agency has sustained its critical function through 51 years and 10 different administrations (6 Republican and 4 Democratic presidents). The change of administration leads to change and priorities, and at times new or rolled back regulations. One such case is the Bush-era fuel standard increasing the fuel standards (established in 1975) to a corporate average of 35 miles per gallon by 2020, further increased by the Obama-era fuel efficiency standard of 54.5 mpg by 2025, then replaced by the Trump-era standard requiring a small 1.5% increase in average fuel economy compared to the pre-Obama standards, and then upon the inauguration of the Biden-era, auto makers themselves began to support higher fuel efficiency standards as the Administration foreshadowed increasing the standards akin to the Obama-era. This presentation aims to provide an overview of current U.S. environmental regulations, where we currently are, where we have come from, and where we are likely to go in the future.

# 628185 Decarbonising Residential Heating: Cost-Benefit Analysis to Support Policy Design

**Abstract:**

The UK's legally binding targets to achieve "net zero" greenhouse gas emissions by the year 2050 may require full decarbonisation of residential heating. This will involve deploying low-carbon heating technologies (such as heat pumps, electric storage heaters, and/or various hydrogen heating technologies) across the country, including in the 24 million homes that currently use natural-gas-based central heating systems. A key challenge of this transition is that the capital and operating costs of low-carbon heating systems are often greater than current natural-gas-based systems. In addition, the UK has a wide variety of building stock, which defies a one-size-fits-all solution.

The aim of this study is to assess the suitability of various policy instruments to promote the uptake of low-carbon heating technologies. A cost-benefit model is being developed based on housing stock data from the English House Survey, performance data from field trials, and cost estimates from the literature. The model will be used to calculate the lifetime cost and levelised cost of heat for a wide range of households and the results analysed to consider how the costs of different low-carbon heating options vary with characteristics such as building type, number of occupants, thermal efficiency, and geographical location. In addition, an uncertainty analysis will used to evaluate the impact of different fuel prices, outdoor temperatures, and personal preferences and to identify the household types likely to be impacted most by these uncertainties.

The results of this study will be presented and discussed in relation to current policies designed to promote low-carbon heating. This study will consider what level of government support will be sufficient to promote low carbon heating over the current natural-gas-based heating systems and will identify if a targeted approach based on geographic location, building type, tenure type, or other characteristics would be more appropriate.

# 625673 Improving the Economics of Industrial Battery Storage: A Proactive Policy and Management Approach

**Abstract:**

Historically, implementation of battery energy storage has been proportionally higher in commercial and residential systems than industrial systems [1]. Economic and operational constraints of industrial facilities make battery installation close to infeasible. Industrial energy usage accounts for about 25% of total energy usage in the US. As the deployment of battery storage increases, lack of implementation in the industrial sector creates a gap, bridged by adapting control and policy strategies to constraints of industrial users. Projections indicate that battery prices will continue to fall, making it even more important to find ways to implement batteries in the industrial sector [2].

This research aims to show that the utilization of electric battery storage to decrease the energy costs and peak power usage of industrial energy users can be made attainable by identifying and closing the policy gap between affordability and utility of industrial battery installations. Similar studies have been done to improve economics and management of storage in residential installations [3]. Other studies have looked at regulatory design of energy storage coupled with renewable sources such as solar and wind [4]. Though these efforts have furthered the penetration of battery storage in the residential and commercial sectors, this same penetration has not been seen in the industrial sector from lack of analysis and specificity to the needs of industry. Policies specific to large electrical users in the industrial sector have also not been explored.

Load data from actual industrial facilities is used spanning a full year of typical manufacturing operations in which the daily electrical peak is about 2200 to 2500 kW. Two models are used, a battery model utilizing the facility load data to look at battery dynamics and effects of system load, and a financial model to analyze overall cost and effect of policy on project economics. The study identifies policies that are currently in place in parts of the US or similar policies that would allow batteries to be financially viable for industrial facilities.

[1] EIA, "Battery Storage in the United States : An Update on Market Trends," Washington, DC, 2020.

[2] W. Cole and A. W. Frazier, "Cost Projections for Utility-Scale Battery Storage," 2019.

[3] M. N. Sheha and K. M. Powell, "An economic and policy case for proactive home energy management systems with photovoltaics and batteries," *Electr. J.*, vol. 32, no. 1, pp. 6–12, 2019, doi: 10.1016/j.tej.2019.01.009.

[4] O. Zinaman, T. Bowen, and A. Aznar, "An overview of behind-the-meter solar-plus-storage regulatory design: Approaches and case studies to inform international applications (Report no. NREL/TP-7A40-75283)," *NREL Rep.*, no. March, pp. 1–76, 2020.

# 621960 Intellectual Property and Carbon Capture

**Abstract:**

As climate change continues to affect the world, industries of all kind are developing new ways to increase their sustainability and decrease their carbon footprint, in part through carbon capture technologies. This green revolution has led to a variety of new proprietary carbon capture technologies being brought to the market. With the increase in new carbon capture technology, the necessity for intellectual property continues to rise. Although companies need to protect their IP rights, there is also a need to allow other companies to help the world become more sustainable. Thus, companies must balance the need to protect IP rights and the need to share their technologies.

This presentation will provide an overview of the basics of IP, how IP is at the forefront of carbon capture technology, and important considerations for members of industry, academics, and anyone with an idea for a carbon capture invention. Speakers will also highlight several case studies addressing sustainable IP and whether the IP was strong enough.

# 632735 Green Technology: Innovation and Regulatory Trends

**Abstract:**

This presentation will discuss current innovation trends and regulatory developments in green technology, which could influence innovation and growth in these industries in the coming years. Green technology is growing and is everywhere around us. It includes electric vehicles, solar cells, programmable thermostats, recycling technology, and many more. According to one source, "[e]vidence of green technology shifting from an alternative investment to a mainstay of the economy is all around us. Around 40% of UK homes and businesses had a smart meter installed by the end of September 2020; on Boxing Day 2020, wind turbines provided the majority (50.7 per cent) of the nation's electricity." California now requires the installation of solar panels on new homes up to three stories high. However, despite the growing prevalence of green technology in the US and abroad, trends show that US patent application filings in this technology have been decreasing for several years. That trend may reverse if governmental investment increases the capital available for innovation.

Recent developments indicate the US government will be investing significantly in the research and development of green technology over the next several years. Indeed, two of the Biden administration's goals are to usher legislation that "makes a historic investment in clean energy and climate research and innovation" and "incentivizes the rapid deployment of clean energy innovations across the economy, especially in communities most impacted by climate change." Innovators in this space stand to gain access to billions of dollars that can fund research and development. This could also mean that many of these innovators' patent rights could be funded by the US government. Accordingly, this presentation will discuss the benefits and drawbacks federal government funding may have on a US patent holder's rights.

The presentation will also include real-life examples illustrating the issues mentioned above. That discussion will focus on examples showing how green innovators and entrepreneurs have taken an idea from concept to patent, achieved funding via their intellectual property, and developed a prototype or pilot plant before heading to market. The presentation will discuss how these strategies can be implemented by researchers, innovators, and entrepreneurs in the coming years in light of the industry trends and potential regulatory changes.

# 639254 Patent Ownership and Inventorship Considerations in Collaborations

**Abstract:**

Inventive breakthroughs often result from collaborative efforts between individuals and organizations across many tech sectors. These collaborative efforts between individuals and entities, as well as the natural movement of employees within their fields, can lead to various inventorship and patent ownership pitfalls for the unwary. This presentation will highlight some of those pitfalls and discuss best practices to mitigate those risks.

# AUTHOR INDEX

**Notes:**

**Notes:**

**Notes:**

**AIChE**
120 Wall Street, FL 23
New York, NY 10005-4020

ISBN 978-1-7138-5717-4

9 781713 857174

CURRAN ASSOCIATES INC.
proceedings.com